Meltdown: Terror at the Top of the World

By Sabrina Shankman

Edited by Susan White

©2015 InsideClimate News
All rights reserved.

ISBN: 0692366032
ISBN 13: 9780692366035
Library of Congress Control Number: 2015901489
InsideClimate News, Brooklyn, NY

"Meltdown: Terror at the Top of the World" was written by Sabrina Shankman and edited by Susan White.

It was produced by InsideClimate News in partnership with VICE.

TABLE OF CONTENTS

A Note to the Reader	v
Chapter One	1
Chapter Two	8
Chapter Three	18
Chapter Four	28
Chapter Five	41
Chapter Six	48
Chapter Seven	50
Chapter Eight	53
Chapter Nine	58
Chapter Ten	73
Marta Chase and Rich Gross	79
Parks Canada	81
The Bears	85
About the Author	87

A NOTE TO THE READER

In the summer of 2013 seven American hikers embarked on a wilderness adventure and came back with a story of terror.

A seaplane dropped them off at the edge of a breathtaking fjord in northern Canada as a frigid rain and an awesome silence fell over them. They were alone in polar bear country and soon to have an encounter for which they were unprepared.

At the top of the Arctic food chain, the polar bear is a magnificent, carnivorous creature. The sea ice where it hunts for seals with ease is fast disappearing, and the open ocean offers it no foothold to secure the fat and meat it needs to thrive. It lives where few people venture, but it is losing its natural habitat to a man-made meltdown, battling starvation and facing extinction.

Those of us reading Meltdown with a full stomach, in the warmth and safety of human civilization, should not feel immune to this tale of terror. The climate change that is thawing the Arctic is upon us all, wherever we may be.

It is not coming by surprise like a wild beast in the dark. It is advancing by our own invitation, drop by melting drop, in parts per million of CO_2 we know how to count, in satellite images we can decipher, in accordance with laws of physics we cannot alter.

Did the seven hikers make it out alive? Will we?

David Sassoon

Publisher

1

With shotguns slung over their backs and radios strapped to their belts, Maria Merkuratsuk and her older brother Eli hiked to a ridge and looked down at the clearing. They were perched at the edge of Nachvak Fjord on the Arctic tundra of Labrador, on the lookout for polar bears.

The view was at once familiar and new. It was the same bountiful land they remembered from their childhood, when their family of 12 spent summers there with other Inuit families, hunting, fishing and foraging during the months when the sea ice melted and allowed their fishing boats to enter the fjord. And yet, it wasn't the same.

The Torngat Mountains, which cut a steep path down to the fjord, had been covered in gray and brown when they were children. But now the landscape was blanketed in green—new growth that has appeared in recent years. Maria Merkuratsuk, 58, swatted occasionally at the flies and mosquitoes that swarmed around her as she walked. Those were new, too.

And the ice—the dependable barrier that bound them to their home in Nain for most of the year, releasing them in the spring to return north, to Nachvak—was no longer something they could count on. It was freezing up *much later and melting much earlier, the warmth of summer lasting longer than ever before.*

As Merkuratsuk absentmindedly grabbed a few blueberries from a low bush at her feet, she saw what appeared to be a large, white rock in the clearing below. In one motion, she dropped to her knee and swung her shotgun into position. As she peered through the scope, the rock began to move. A polar bear.

In all her years here as a child, she had seen only a handful of polar bears. Now she rarely ventured into Nachvak Fjord without seeing one.

Eli Merkuratsuk headed down to scare the bear away, firing flares as the animal ambled on all fours into some tall willows nearby. The idea wasn't to harass the bear, but to make sure it learned to fear interacting with humans. It was one of many ways he did his job as a bear guard. As the Merkuratsuks knew, they were just visitors here. This land, this fjord, belonged to the bears.

Eli Merkuratsuk drew closer and the bear—large, with a thick, white coat—left the willows and walked toward him. Instead of running away, it rose onto its hind legs and stood there a moment, showing its height and towering over him. Then it dropped down onto all fours and walked slowly back into the bush.

The ad in a fall 2012 issue of Sierra magazine promised the adventure of a lifetime: two weeks trekking through the untouched lower reaches of Canada's Arctic tundra, with the possibility of seeing the world's largest land carnivore, the polar bear.

Participants must be fit and experienced hikers, the ad warned. They would also have to accept an element of risk, including lack of access to emergency medical care. But the payoff would be big.

"If you dream of experiencing a place that is both pristine and magical, a land of spirits and polar bears rarely seen by humans, this is the trip you have been waiting for," the ad said.

Two seasoned Sierra Club guides, Rich Gross and Marta Chase, would be leading the trip. Gross, now 61, worked for a low-income housing nonprofit in San Francisco but since 1990 had spent a week or two each year guiding Sierra Club trips in remote parts of the world. With short, curly gray hair and an easy smile, Gross lived for these adventures. Chase, 60, was a medical diagnostics consultant who'd been leading hiking trips since she was in high school. She and Gross had guided 14 trips together.

It was Gross's idea to go into the Torngats, one of Canada's newest national parks. He'd never seen a polar bear in the wild and was drawn to the spiritual appeal of the mountains. The park was named after Torngarsuk, an ancient Inuit spirit that appeared as a polar bear and controlled the lives of sea animals. The terrain itself has a mystical appearance, with sharply peaked mountains and fjords cutting into the land from the coast of the Labrador Sea. Only a few hundred people venture there each year, and Gross wanted to be part of that exclusive group.

Chase wanted to see the park, too. But she worried about hiking in polar bear country.

Polar bears sit at the top of the Arctic food chain. A large male can weigh as much as 1,700 pounds and stand 10 feet tall. Unlike black bears or grizzlies, polar bears are carnivores through and through—they can't survive without meat. They live most of their lives on sea ice, lurking near holes, watching and waiting—sometimes for hours, sometimes for days—for their favorite prey, the ring seal. When a seal surfaces to breathe, the bear pounces, grabbing it by its head and crushing its skull.

Polar bears typically stay clear of humans. But if there were a time and a place to see one, the Torngats in mid-summer would be a good bet. That's when the sea ice melts for a while, forcing the bears onto land.

To ease her concerns, Chase studied websites, talked to experts and read books. She knew Gross was obsessive about safety, but both of them were responsible for keeping the group safe and she wanted to be prepared, too.

In New York City, 65-year-old Larry Rodman signed up the same day he saw the ad on the Sierra Club website. The walls of Rodman's midtown Manhattan law office were adorned with photos of wildlife and scenery he'd taken on past wilderness trips, including one with Chase and Gross. But he'd never seen a polar bear. The timing of the trip was especially good, because he was pondering a big life change. He had decided to leave his law partnership and apply to Yale's environmental policy program.

Marilyn Frankel, a 66-year-old exercise physiologist from West Lynn, Ore., had traveled with Gross and Chase many times before, and they were close friends. After some years away from backpacking, this would be her first trip back—and one of the most remote and intense of her life.

Rick Isenberg, a 56-year-old Scottsdale, Ariz. physician who does clinical research and regulatory work, signed up because he wanted to get away—not to be alone, necessarily, but to appreciate being small in the vastness of wilderness. This would be his third Sierra Club trip but by far the most adventurous and extreme. He hired a personal trainer and began taking long hikes in the Arizona heat.

Matt Dyer, a 49-year-old legal aid attorney in Turner, Maine, was dreaming of adventure when he filled out the forms for the trip. He wouldn't have considered spending so much money—$6,000—on such a luxury, but he was at a point in his life where he could afford it. He'd always wanted to go to Labrador, because his family roots traced back to the province. And getting to Montreal, the starting point, wouldn't be too expensive.

But Gross was concerned about Dyer's lack of experience.

"This trip requires backpacking experience and I don't see any on your forms," Gross said in an email to Dyer. "This is a particularly tough trip since it is all off trail and packs will be quite heavy (50+ pounds). The area is remote and evacuation is only by helicopter."

Dyer told Gross he was in good shape and had been hiking and camping in New England for years, including some trekking with the Appalachian Mountain Club.

"I'm not a city person (I grew up on an island about 8 miles from the mainland) so being away from the [7-Eleven] is not going to bother me," Dyer wrote. "I totally understand that you don't want to wind up a thousand miles from nowhere with a problem, but I think I can do this."

Dyer agreed to follow a strict training plan, and Gross agreed to take him.

As the hikers prepared for their journey in the fall of 2012, University of Alberta and long-time Environment Canada biologist Ian Stirling had just seen the publication of his most recent paper, a review of dozens of scientific reports that explored how climate change was affecting polar bears.

Stirling has devoted 40 years of his life to studying polar bears, and many consider him one of the preeminent biologists on the subject. His long-term studies in Canada's western Hudson Bay had helped establish the link between polar bears and climate change. That link is now so strong that the bears have become one of the most visible symbols of global warming.

Stirling had written this new paper for the journal Global Change Biology with Andrew Derocher, one of his former PhD students who is also a leader in the field. The long-term picture they painted is bleak.

Temperatures are climbing faster in the Arctic than anywhere else in the world, leading to a substantial decrease in sea ice, where ring seals and the bears' other prey live. In September 2012, the Arctic sea ice level was 49 percent lower than the historical average from 1979-2000.

The southern parts of the Arctic, including the Torngats, have had an ice-free summer season throughout modern times. But the ice-free period is growing longer. Since the late 1970s, the number of ice-free days in the area around the Torngats has increased from 125 days to 175 days.

In the not-so-far-away future, the Arctic could look much different than it does now. In a worst-case scenario in which carbon emissions continue to rise, midsummer is projected to be completely ice-free by the middle of this century.

"If the climate continues to warm and eliminate sea ice as predicted, polar bears will largely disappear from the southern portions of their range by mid-century," Stirling and Derocher wrote in their paper. Bears would probably survive in the northern Canadian Arctic

Islands and northern Greenland for the foreseeable future, but their long-term viability, they wrote, "is uncertain."

The scenario they laid out was straightforward.

Less sea ice means polar bears must spend more time on land, where their specialized hunting skills are useless. To survive, they live off the body fat stored from their earlier kills on the ice. As the period when they have to live off that reserve grows longer, some eat goose eggs, grasses or berries. But their foraging goes only so far—they can't survive without the fat they get from seals.

"As the bears' body condition declines, more seek alternate food sources so the frequency of conflicts between bears and humans increases," the scientists concluded.

After all, to a starving bear, a human is just meat.

2

Matt Dyer lugged his 50-pound pack into the Quality Hotel Dorval in Montreal on July 18, 2013. To save money, he'd taken a 12-hour overnight bus from Lewiston, Maine, and then spent the morning wandering around Montreal. He ate two breakfasts and killed time by napping in a park, feeling "kind of like a bum." By the time he could check in, the afternoon sun was hot, and he was tired.

Larry Rodman walked in at the same time, fresh off the airport shuttle bus after a quick flight from New York City.

The two men dumped their bags, got some lunch and started talking. Rodman, the big city law partner, and Dyer, the legal aid attorney with the scraggly gray ponytail, hit it off immediately. For Dyer, especially, that was a relief. He'd been less concerned about the arduous journey than about the people he'd be trapped with in the wilderness. When you're paying for the trip of a lifetime, you want to enjoy the company.

Gross and Chase had flown in a day earlier to buy supplies and make last-minute arrangements.

Chase's husband, Kicab Castañeda-Mendez was there, too. Chase had introduced him to hiking when they were in their 30s and since then the 64-year-old management consultant had joined all of her Sierra Club trips, pitching in with the planning and playing the default role of group photographer. On this trip his presence was even more reassuring. In June, Chase had been diagnosed with breast cancer and was told she needed a mastectomy.

Petite and fit, Chase was shocked by the news. At first, she assumed she'd have to cancel the trip. The idea of backing out compounded the blow of the diagnosis. But when her doctor assured her that postponing the surgery for a few weeks wouldn't matter, she decided to go.

Over the years, Chase and Gross had developed a division of labor for their trips. Chase handled logistics like transportation and meals. Gross handled park permits and the route, poring over maps and plotting out the various options for getting from A to B. Each year they alternated who took the lead on research and outreach. For the Torngats trip, it was Chase's turn, although Gross continued to read up on staying safe in polar bear country.

Chase studied the website for the Torngat Mountains Base Camp & Research Station, which Canada's parks department had opened in 2006 in the Nunatsiavut region. This autonomous area in Labrador, includes five small Inuit communities plus Torngat Mountains National Park. In 2009, the Canadian government transferred ownership of the Base Camp to the Nunatsiavut government. The camp sits just outside the southern limit of the park, on Saglek Fjord.

Chase sent an email to Base Camp, thinking it might serve as their entry point to the park. When she didn't get a response, she contacted Vicki Storey, an adventure travel agent in Alberta who'd been booking travel to the Torngats since before it became a park. Storey sent Chase to Alain Lagacé, who operated two camps that offered guided tours, fishing expeditions and wildlife safaris.

Lagacé knew the area well, Storey assured Chase. He'd been arranging trips into the Torngats for decades. He was also a good choice for a trip like the ones Chase and Gross specialized in, minimalist expeditions in which they acted as their own guides. There would be no planned excursions or activities on this trip, just unadulterated backpacking.

In emails and phone calls over the course of months, Chase and Lagacé shaped the trip. The group would fly from Montreal to Kuujjuaq, the largest Inuit community in Nunavik, the Inuit region of Quebec. Then they'd take a small charter plane to Lagacé's Barnoin River Camp and spend the night. The next morning a floatplane would deposit them in the Torngats, where they'd be on their own for 11 days.

"The thought of polar bears is still a concern to me," Chase said in one of her emails to Lagacé. "I have experience with black and brown bears but not with polar."

Guns are generally prohibited in Canada's national parks, but in 2011 Parks Canada broadened the rules for parks with polar bears. Researchers, guides licensed by Parks Canada, and local, native Canadians were allowed to apply for gun permits for those parks. Guns could also be carried by Inuit "bear guards" who have taken a polar bear safety course and been licensed by Parks Canada.

"Polar bears are present in ten of Parks Canada's northern national parks and there is a documented history of human-bear interactions in many parks," Parks Canada said in a document explaining why the regulations had been loosened. "The impacts of climate change on sea ice may result in changes in the density and behavior of polar bears on land in these parks. These factors bring an increased risk of dangerous human-bear encounters."

For visitors who didn't qualify for a gun permit, Parks Canada's website for Torngat Mountains National Park "strongly" encouraged—but did not require—hiring a bear guard.

"The accompaniment of a bear guard will allow you to relax and enjoy your hike, but will also give you the opportunity to experience the park with the help and guidance of Inuit who truly know the land," the website said.

The Base Camp website also addressed the issue of guns and bear guards:

"Visitors should note that it is illegal to carry a firearm in the national park. You are required to be familiar with the use of bear deterrents, and to bring approved deterrents with you. Alternatively, we recommend engaging the services of an Inuit guide for the duration of your visit. Inuit guides are permitted to carry firearms in the park and are well-trained in visitor safety procedures."

When Chase asked Lagacé whether they'd need a bear guard, she said he told her that no one who traveled through his camp used bear guards, and that there had been no problems, provided they took other precautions.

"Regarding the safety against polar bears, we have it all," he wrote. "The 12 gauge magnesium frare (sic) gun are working extremely well, plus we have the pepper spray, and the pepper spray grenade (sic) and electric fence. These have worked very well in the past but there are always precautions to be taken. Never cook food in your tent, don't leave trash around your camp site, avoid camping along the shore of a coastal lake, etc."

Chase and Gross decided that flare guns, bear spray and electric fences would offer them the protection they needed. Although Chase still had a sense of foreboding about the bears, she felt they'd done their homework and would be safe. Lagacé clearly knew what he was talking about. And none of the Parks Canada employees she and Gross had talked with had even mentioned bear guards, she said, let alone recommended that they use one.

Chase and Gross arranged to rent two flare guns from Lagacé. They would carry four shells for each gun.

Gross picked up two electric fences from the Sierra Club—one to encircle their campsite, the other to protect the area where they would cook and store their food.

The instructions were missing, and the pieces were wrapped haphazardly. So Gross asked a close friend who is an electrician to help him practice setting them up outside his house in San Francisco.

Because the fences hadn't come with any instructions, Gross didn't see the warning they contain: "This product is a bear deterrent which may protect users in some unexpected confrontations with bears but may not be effective in all situations or prevent all injuries or damages."

Each fence stood about three feet high and consisted of three parallel wires suspended from four-foot posts. Although the wires looked flimsy, they carried five to seven kilovolts of charge—not enough to seriously injure a bear, but supposedly enough to send it running.

Gross emailed a picture of the fence in his front yard to Castañeda-Mendez.

"What's the polar bear supposed to do? Die of laughter?" Castañeda-Mendez wrote back.

They'd also take bear spray, which they had carried on previous trips to ward off grizzly bears. Should something go wrong, they would have a satellite phone to call for help.

For more information, Lagacé told them to contact Parks Canada, which manages Torngat Mountains National Park.

A Parks Canada employee told Chase that anyone entering the park was required to watch a DVD on polar bear safety. Parks Canada agreed to send the video to Lagacé's camp, so they could watch it right before they entered the park. The employee assured them they were in good hands with Alain Lagacé.

Parks Canada also discussed options for their backpacking route. They decided to start at Nachvak (pronounced *Nock-vock*) Fjord and move inland, packing up camp most mornings and working their way toward Komaktorvik Fjord, where they'd be met by a plane from Lagacé's camp. It would be a tough trip. They'd be carrying their 50-plus pound bags for about six hours each day, stopping in mid-afternoon to find a place to camp. Halfway through the trip, Lagacé would send a plane to drop off the rest of their food.

As the group gathered in Chase and Castañeda-Mendez's hotel room in Montreal on July 18, Gross and Chase felt confident about their preparations. They double-checked each hiker's gear and divided their food supplies into baggies labeled breakfast, lunch and dinner. Then they packed the plastic bags into bear canisters—portable lockers designed to keep food from attracting unwanted attention while they hiked—and stuffed them into the hikers' packs.

The conversation about climate change and its consequences often revolves around abstract concepts—sea level rise, ocean acidification—but that's not the case with the melting of the sea ice. In the Arctic, the consequences are more tangible, more immediate.

Using yearly averages taken throughout the Arctic, NASA scientist Claire Parkinson has reported that about 695,000 square miles of sea ice have been lost since 1979, when satellites began recording images of the region. To put that into context, that's roughly the same as if the western portion of the United States—California, Nevada, Oregon, Washington, Arizona, Utah and most of Idaho—had disappeared.

The Arctic's thick white expanse of sea ice resembles land but for the fact that it moves with the ocean below it, shifting along fracture points called leads. In the northern regions of the Arctic, the ice gets thicker and stronger from year to year. In the southern regions, it melts in the spring and then builds up in the fall. Initially it appears like a greasy stain on the ocean's surface, quickly growing to a foot thick, often in just a matter of days. In the course of a single week, open ocean becomes a hard surface, a habitat capable of sustaining an entire ecosystem.

Parkinson specializes in sea ice at NASA's Cryospheric Sciences Laboratory, where she is a climate change senior scientist. She explains that sea ice has a symbiotic relationship with climate change. It's not just that the ice is melting—but also that its disappearance is exposing the dark ocean below. A surface that once reflected the sun's radiation is being replaced by a surface that absorbs it, further warming the ocean and leading to even more sea ice melt.

This process, called "ice albedo feedback," contributes to a phenomenon called "polar amplification." It refers to the increased rate of warming near the poles in response to rising temperatures, which are precipitated by greenhouse gas emissions.

In a study in Geophysical Research Letters, published in June of this year, Parkinson analyzed the satellite record and found that since 1979 there has been an average of at least five fewer days of sea ice per decade in areas with seasonal ice.

In some areas, the decline is much steeper.

Parkinson looked at the number of ice-free days in the Davis Strait, which is part of the eco-region that includes Torngat Mountains National Park. She found a decrease of about 15 days per decade—or roughly 50 days since 1979.

Reports by the Intergovernmental Panel on Climate Change (IPCC) earlier this year also quantified the sea ice loss. It said that Arctic sea ice has disappeared at a mean rate of between 173,000 and 196,000 square miles per decade since 1979—a loss larger than the state of California every 10 years. The ice is disappearing even faster in the more southern areas of the Arctic—between 280,000 square miles (California plus Arizona) and 410,000 square miles (California, Arizona and Colorado) per decade.

It wasn't long ago that scientists who came out with such alarming findings faced skepticism and ridicule.

In 2006, Cecilia Bitz co-authored an article in the journal Geophysical Research Letters that projected the Arctic would have its first completely ice-free period by the end of the summer of 2040. Bitz, a physicist who studies sea ice and does climate modeling at the University of Washington, said some media panned the paper as the work of "these crazy climate scientists." The Village Voice ran a cartoon mocking the findings.

Even Bitz had trouble internalizing the magnitude of what she had learned.

In the summer of 2007, she was teaching at the International Polar Year Summer School in Svalbard, which is Norway's Arctic archipelago. Since she does her climate modeling work from her office, it was her first visit to the Arctic.

In retrospect, she says she should have been shocked by the unseasonably warm weather she encountered and by the fact that she could go for a run wearing shorts. But when the readings of 2007's sea ice extent came out, she, like many of her colleagues, was caught by surprise. By mid-August of that year, the sea ice minimum had broken every existing record—and there was still a month to go before it hit the annual low-point.

By the time the ice melted to its minimum that year, it was almost 40 percent below the 1979-2000 average.

"To be that fooled by what came to pass was really shocking to me and a big wake-up call," Bitz said. "I think the whole community felt that way. We were startled."

The media called on Bitz again, this time requesting interviews that could help put the record low in context. For Bitz, it could have been redemptive, were the facts not so disheartening.

In the years since, Bitz has been part of a worldwide effort to improve sea ice modeling. When the next record-breaking low came around, in 2012, Bitz and other experts were less shocked.

Another 300,000 square miles of sea ice had been lost—more than the state of Texas.

Sea ice melts at different rates each year. Sometimes the pace picks up, sometimes it slows. But Bitz says the key isn't what happens in an individual year. What matters is the trend—think climate, not weather—which has been persistently downward in the past few decades.

As the melting continues, a ripple of change reverberates through the flora and fauna that rely on sea ice as their habitat—the capelin and other fish that harvest the plankton along the ice's edge, the narwhals and beluga whales that swim below, and, of course, the polar bear, the king of the Arctic, sitting patiently on the shrinking ice, waiting for its prey.

3

On the afternoon of July 19, the 19-seat Air Inuit Twin Otter descended over a steep waterfall and bumped down onto the gravel landing strip at Barnoin River Camp, some 900 miles north of Montreal. As they stepped off the plane, the Sierra Club backpackers walked into a wall of mosquitoes—a swarm that would harass them throughout their stay.

Although they wouldn't head into the wilderness until the next morning, in many ways it felt like they were already there. The camp is a series of small plywood structures set on concrete blocks near the banks of the Barnoin River. The water is so clear they could see trout swimming under the surface. The river also serves as a driveway for floatplanes. With no roads, planes and helicopters are the only way in or out of the area.

Lagacé, a fit, middle-aged man with a gray mustache, gave them an orientation, pointing out the bathrooms, kitchen, dining facility and bunkhouses. The group paired off, deciding who would sleep where. The two lawyers—Rodman from New York City and Dyer from Maine—bunked together. Already they felt like old friends.

Gross, Castañeda-Mendez and Isenberg unwrapped the electric fences they got from the Sierra Club, to give them a final test.

They were manufactured by a company called UDAP, which describes them on its website as "an answer to the problem of fear while camping in bear country." In a testimonial, a backpacker describes waking up in his tent in the Torngat Mountains, roused by "a loud chuffing sound" from a polar bear outside.

"The mother bear was just outside the fence, no more than 12-15 feet away from me, looking at me warily and looking back at her cub," he wrote. "Slowly, she walked away and herded her cub up a slope just outside the fence. It was clear to me that the mother bear had approached my camp and touched the fence right near the red energizer and the shock had caused her to make the loud chuff in surprise ... After a few minutes the mother and cub walked behind a large boulder and disappeared."

When the men looked at the fences, what they saw wasn't particularly inspiring. So Lagacé offered them two of his own fences, which were in much better shape. They came with directions, were wrapped properly and looked sturdy.

The men set up the fences one at a time. As they worked, they swatted at the black flies and mosquitoes. Bug spray—even the strong stuff, with 100 percent Deet—didn't make much difference.

They switched the first fence on, but couldn't tell if it was working. No one volunteered to touch it and find out.

Lagacé came to the rescue with the voltage tester he uses at his camp. A small rectangular device with a number of lights, it's attached to a thin pole that is pushed into the earth to ground the device. When the device is touched to the fence, the lights are supposed to come on. As the voltage increases, more lights illuminate.

When Lagacé touched the tester to the fence, only a few of the lights lit up. Castañeda-Mendez got the same result. Gross wasn't surprised. The voltage tester they were using wasn't mean to read this type of fence, which sends pulses of electricity, rather than a constant current. But at least the test proved that some electricity was pulsing through the fence.

They broke down the first fence and set up the second, with the same results. Most of the hikers were reassured. But Isenberg, the doctor, still had reservations.

As the others stood around talking, he thought, *the heck with this*, and grabbed the fence with his bare hand.

Instead of a shock, he felt a light tingle.

"Look at your feet," Lagacé said, pointing to Isenberg's hiking boots. They had rubber soles, perfect insulation for that kind of shock.

Isenberg decided not to try again without his shoes. He still had some reservations, but he was ready to put his faith in the fence.

Swatting at the bugs, Gross pulled out one of the flare guns. He'd never shot one before and wanted to be comfortable with the way that it handled and to see how far it would shoot.

The bright orange gun looked almost like a toy. Gross pulled the trigger, and the 12-gauge flare shell erupted with a bang. The bright light of the flare shot about 150 feet in a straight path toward the ground. Upon impact, the cartridge exploded with a *whoosh* and a second burst of light.

Frankel saw the flashes from a shed where she was sorting food, pulling out the half that would be airdropped to them midway through the hike. *OK*, she thought to herself, *those should work.*

Later in the afternoon, the cloud cover thickened, with temperatures in the 40s. Rodman and Dyer took cover in their bunkhouse, napping and chatting. They found they had a lot in common—both had fenced and both loved opera. Rodman told Dyer about his decision to leave his law practice and go back to school. Dyer told Rodman about growing up on Cliff Island, Maine, worlds away from New York City. When Dyer changed his shirt, Rodman got a glimpse of the tattoos that covered Dyer's shoulders and back, all images from nature—a turtle, a winged bull, a giant tree of life with ravens.

At dinnertime they headed to one of the main buildings to eat and watch the Parks Canada DVD on bear safety. The seven backpackers said Lagacé told them the DVD hadn't arrived, so they asked him to talk to them about safety in polar bear country. Lagacé disputes that account; he says he showed them the DVD.

As they dug into their dinners, Lagacé shared stories about his run-ins with polar bears and what he had learned in his decades of bringing people into the Torngats.

The most important rule, he said, was to be aware and prepared at all times. Polar bears aren't like the grizzlies they were familiar with, he warned—they're hunters, looking for meat. Because the bears travel along water, the group should camp away from the edge of the fjord. If they obeyed those rules, and slept within the perimeter of their electric fence, he said they should be just fine.

They headed to their cabins that night with Lagacé's warnings fresh in their minds. But mostly they were just excited for the day to come, when their adventure would begin.

In February 2013, five months before the Sierra Club group entered the Torngats, more than 35 polar bear scientists, conservationists, tour operators, government and community representatives and police gathered in Tromsø, Norway, to discuss an increasingly relevant question: What do we know about polar bear attacks on humans—and are such attacks on the rise?

The participants came largely from the "range states" where polar bears are found—Canada, Greenland, Norway, Russia and the United States—as well as Denmark, the United Kingdom and the Netherlands. When humans and bears collide, the people at that conference would likely be the first responders.

One of the key speakers was James Wilder, who at the time was a U.S. Fish and Wildlife Service biologist. He was compiling a database that tracks the number of attacks and catalogues key information: the bears' health, what the humans did to deter the attack, and whether anyone died. Wilder's data are still incomplete. Some of the range states haven't submitted their reports, and even those that have don't always have a complete accounting of the incidents.

But two things are already clear.

The number of people killed by polar bears is relatively small. So far, Wilder had found just 21 deaths in the last 140 years.

But the number of interactions between humans and polar bears is rising.

According to Wilder's tabulations, there were fewer than 10 attacks per decade in the 1960s and 1970s. But in the first four years of this decade, Wilder has already documented 14 interactions. At this pace, he expects to see about 35 incidents by the end of 2019—nearly as many as the last 40 years combined.

The run-ins Wilder has documented are often well publicized. The media love stories about polar bear attacks, and the viral nature of the stories makes it seem as if they happen more often than they do. But the stories spread for a reason: the attacks are highly dramatic.

In 2011 a cell phone camera captured a bear biting and pawing a woman in the middle of a northern Russian town. In 2013, a 40-year-old man was chased down a main street and pinned in the doorway of a bakery in Churchill, Manitoba; he scared the bear off with an illuminated cellphone screen. A month later, in the same town, a 30-year-old woman was attacked while she walked home from a Halloween party at 5 a.m.

Some incidents slip through the cracks.

One that doesn't appear in Wilder's records took place in 2009, not far from the Nachvak Fjord in Torngat Mountains National Park, where the Sierra Club group was headed.

A group of hikers had arrived by boat at the North Arm of the Torngat Mountains for a camping trip. Before they even pitched their tents, a polar bear swam up to the shore and approached them. Their bear guard, John Merkuratsuk, followed the standard protocol, gathering the group together and making loud noises. One group member fired flares.

But the bear kept coming.

Merkuratsuk loaded his gun with bullets, but it jammed. With the bear moving closer, Merkuratsuk cleared the gun, reloaded and fired. He shot the bear three times before it died.

In photos the group took as the incident was unfolding, the bear looked as scrawny as a street dog, its ribs protruding and its head appearing unnaturally large. When some of the Inuit at Base Camp skinned the bear for its pelt, they found that it was severely underweight and had an abscessed tooth that could have contributed to its poor health and its bold attack.

Other incidents in the Torngats are also missing from Wilder's log.

Alfred Duller, a 63-year-old retired schoolteacher from Austria who traveled in the park for about 30 years, said bears pulled him out of his tent in the Torngats at least three times. Although the details of some of those attacks are known by people who live and work near the park, they don't appear in Wilder's data.

In one incident, Duller was sleeping on the shore of Ikkudliayuk Fjord, not far from the northern tip of the Torngat Mountains, with his shotgun zipped into his sleeping bag. Suddenly the tent was flattened on top of him and a bear bit into the hood of his sleeping bag. Duller crouched in the bottom of the bag, unable to reach his gun. He yelled to a friend sleeping in a tent nearby and the friend screamed, "Flatten!" Duller scrunched down in the bottom of his sleeping bag while his friend fired nine shots into the bear.

The men were compassionate, experienced travelers, with great respect for polar bears, and they wanted to make sure the animal died quickly.

Duller stopped visiting the Torngats when the mountains became a national park and gun permits were granted only to Inuits, researchers and licensed guides. To go there without a gun, he said, would be "suicide."

Wilder, the U.S. Fish and Wildlife researcher, can't say with scientific certainty why the number of incidents is increasing. Maybe it's because more people are traveling into polar bear country. Or maybe it's because the melting sea ice is forcing bears to spend more time on land, away from the ringed seals that are their primary prey.

According to Wilder's data, 70 percent of the bears involved in fatal attacks on humans were in below-average body condition, meaning they were skinny or thin. Sixty-three percent of the bears in the non-fatal attacks fell into that category.

"Obviously that's a concern if sea ice is melting and bears have less access to their normal prey," Wilder said in a recent interview. "So they're in poorer body condition and they wind up on shore because the ice melts. That's a worry for people living along the coasts in polar bear country."

It's unclear whether the bears' health played a role in two vicious attacks that Norwegian police officers described on the third day of the Tromsø conference.

In the first, in 2010, two Norwegian kayakers were trying to paddle the 1,250 miles around the Norwegian archipelago of Svalbard. They had camped in an inlet on the island of Nordaustalendet and were sleeping in a two-man tent surrounded by a trip wire connected to a flare.

In the early hours of the morning, a polar bear reached into the tent and clamped its jaws around the head of 23-year-old Sebastian

Plur Nilssen. As the bear carried Nilssen away, it punctured his lung, head and neck, narrowly missing an artery. The other kayaker shot and killed the bear. Nilssen was airlifted to a hospital and survived.

The trip wire and flare had failed. The bear was an adult male that weighed 784 pounds—on the low side of average—and had no existing injuries or disease. The probable cause in Wilder's database is listed as, "Predatory on human - tent."

A 2011 encounter ended more tragically. A group of British students and their guides camped near the Von Post glacier on the island of Spitsbergen, which is also part of the Svalbard archipelago. Their tents were protected by a trip wire attached to explosives that were supposed to detonate if triggered by a bear.

At about 7:30 a.m., a polar bear tore a hole through the wall of 17-year-old Horatio Chapple's tent and grabbed the sleeping teenager by the head. Horatio's screams woke the others, who scrambled from their tents in time to see the bear rear up and slam the teenager to the ground. While the group leaders wrestled with what turned out to be faulty rifles, two other students and two of the leaders were mauled. The leaders finally managed to shoot and kill the bear. Horatio died from his injuries; the others survived.

The bear that killed Chapple was old and weighed just 551 pounds, about half the average weight of an adult male polar bear. It had worn and probably painful teeth that could have left it starving.

The Norwegian officers ended their talk with a warning: Hikers shouldn't rely solely on fences to keep them safe in polar bear country. In both of the cases they described, the campers had so much confidence in their fences that they hadn't posted overnight watches. When the fences failed, the campers in their tents were like fish in a barrel for polar bears. Or, perhaps more aptly, like seals in the sea ice.

When the Sierra Club hikers woke at Barnoin River Camp on July 20, they discovered that clouds had socked them in and the floatplane that was supposed to take them into the Torngats couldn't take off.

Unfazed, they spent the day climbing in the mountains that surrounded the camp, their bright orange, red and yellow jackets standing out in the stark landscape. The mosquitoes were relentless. Chase covered her hair with a kerchief to keep them away. Dyer tucked a pesticide-laced bandanna under his baseball cap to protect his neck.

They hiked to Barnoin Waterfall, which they had seen when they flew in. It has a vertical drop of 248 feet, and at the point where it crests, the dark water turns a light, iridescent shade of green before crashing to the river below. The sound of rushing water echoed through the area.

As the weather cleared, the temperature rose and they took off their jackets before heading back down. Pockets of snow looked like white puddles in the green mountaintops.

Their arrangement with Lagacé included only one meal, so that evening they prepared what had been on the menu for their first night in the wilderness: tomato soup, pasta with spinach and cheese sauce and chocolate pudding. Frankel noticed a natural cohesion to the group that she hadn't always seen on past trips. Without direction, each person pitched in and took on different responsibilities.

It was a small thing, but to Frankel it signaled that they'd be in good shape for whatever came their way when they were alone in the mountains.

4

On Sunday, July 21, a floatplane carried them over the western portion of the Torngat Mountains and then descended toward the eastern shore, weaving between the final peaks. The landscape they saw on the 45-minute flight was desolate but breathtaking—treeless, with ice covering parts of the glassy lakes below. Rivulets of icy water cascaded from mountain peaks that jutted into the cloud-filled sky.

The plane landed flawlessly on Nachvak Fjord, backing into the shore so they could exit without getting their feet wet. It was the kind of impressive maneuvering that comes from using planes the way suburban commuters use cars.

Castañeda-Mendez held onto the plane's pontoon while the others off-loaded their gear. The pilot said his goodbyes and the sound of the engines receded into the distance, leaving them alone with just the sound of small waves on the shore. The skies were clear, but a cold rain started to fall.

In a place like this, untouched as it was, it was easy to imagine a world before humans. Fjords are usually formed when the ocean pours into valleys left behind by melting glaciers. What was left behind on Nachvak Fjord felt prehistoric—like the end of the Earth, with the long fingers of the fjord reaching into the shoreline.

Chase and Gross left the group on the shore while they scouted for a suitable place to set up camp. Lagacé had warned them not to pitch their tents on the beach. He said they should find a high place to sleep because polar bears are known to come right up the fjord where they had landed.

But when Chase and Gross reached an area that met Lagacé's criteria—an elevated spot about a quarter mile away—they discovered it didn't have easy access to drinking water. Further down, a bit closer to where they had been dropped off, they found a spot that looked ideal: flat enough for comfortable sleeping and cooking, with easy access to fresh water. It was still at least 150 yards away from the shore, and people had obviously camped there before. They'd left behind stakes and piles of rocks.

On the shore, the rest of the group was jubilant. To mark their official entrance into the wilderness, they took pictures of each other. As if on cue, a rainbow appeared. Dyer took it as a good omen.

The first thing they did at the campsite was set up their electric fences, using rocks to help stabilize the poles. They wound the wires around the poles, making sure everything was taut.

The perimeter of the sleeping area was 27 feet by 27 feet, a bit larger than a boxing ring. It wasn't a big space, but it was as far as the fence could stretch. The six tents would be separated by about three feet. Each person had his or her own tent, while Chase and Castañeda-Mendez shared a slightly larger tent.

On every trip, Rodman made a point of finding out who the loudest snorers were, so he could pitch his tent as far away as possible. In this case, he didn't need to ask around. His new friend Dyer was a serious snorer. Once Dyer set up his tent, Rodman picked a spot in the opposite corner.

They set up their food and cook station inside the second fence, about 200 yards away. The cook tent was shaped like a tepee and could be adjusted to sit a few feet above the ground so air could circulate while they cooked. With blue and gray triangular panels, it looked like a small circus tent.

Once everything was ready, they flipped on the switches for both fences and watched the lights on the battery packs flicker to life. They felt secure, protected—and they were pleased to see that they were in a prime location for viewing wildlife.

"No sooner had we gotten the tents up then we looked down towards the water a ways away and there was a wolf," Chase said.

From the smallest plankton to the largest marine mammals—whales and polar bears—there's a chain of connectedness on the Arctic tundra that allows nothing to operate in a vacuum. All the disparate species share the same dynamic ecosystem, and all of them rely on the same habitat: the ice.

The food chain that leads to polar bears starts with phytoplankton, tiny, free-floating, plant-like organisms that live in water and in the ice. In the spring, the seasonal sea ice break-up triggers a phytoplankton bloom, and light green shelves of it swirl into the Arctic Ocean, signaling that winter's cold grip is fading, at least for a while.

In recent years, however, these blooms have been showing up in places where they haven't been seen before. In 2011, scientists found whole swaths of the Arctic teeming with phytoplankton blooms.

What's happening, says physicist Cecilia Bitz, is that old, thick sea ice is being thinned by the warmer ocean temperatures. That allows more sunlight to permeate the ice surface, stimulating the phytoplankton to grow within the ice. Because the blooms are much darker than ice or snow and absorb more energy from the sun, they trigger further melting.

In the southern parts of the Arctic, where these blooms have always been part of the ecosystem, the spring melting is happening earlier. That can trigger an earlier bloom, which in turn sets off ripples that affect zooplankton—miniscule, free-drifting organisms, like shrimp larvae or tiny marine bugs, that feed on the phytoplankton. That affects fish like capelin, a small fish from the smelt family that feeds on the zooplankton.

Historically, this has occurred in spring, when the capelin feast on the bloom along the shelf of the retreating ice. The plankton is high in fat, and the fish rely on it for the growth of their reproductive systems. As the ice melts earlier, however, the timing is being thrown off. The zooplankton isn't getting to the phytoplankton on time, and the fish don't have as much to eat.

A 2014 study in the peer-reviewed journal *PLOS-One* found that as ice retreats earlier, capelin numbers drop off. That leaves the species at the next level of the food chain—seals—with less access to one of their primary food sources.

There are many types of seals in the Arctic, but bears in the Davis Strait and the Torngats thrive on harp seals, whose scientific name reflects their reliance on ice: *Pagophilus groenlandicus*, meaning "ice-lover from Greenland."

Each spring, harp seals follow the ice break-up south, to Labrador, Newfoundland and the Gulf of St. Lawrence, where they haul out onto the ice and give birth to pups, which can't yet swim. Polar bears flock to the seals' whelping patches, because the pups make for high-calorie, easy-to-catch meals at a crucial time, when the bears need to put on weight before the ice-free months.

In some areas, however, the supply of seal pups is declining.

Garry Stenson, head of the marine mammal section at Canada's Department of Fisheries and Oceans, has studied harp seals for decades and oversees a pup population survey every few years. He has found that pup survival declines drastically in low sea ice years. In 2011, 70 percent of the harp seals born in the northwest Atlantic did not survive. The lower numbers are attributed in part to higher rates of miscarriages, called "late-term abortions" for seals. A 2013 study connected the miscarriages to lower biomass among capelin in years with less sea ice.

"These ecosystem changes are predicted to continue," Stenson wrote in a 2014 assessment of northwest Atlantic harp seals. "Therefore, it is likely that reproductive rates will remain low."

The world's harp seal population is so abundant that it can bounce back between low-ice years. But that doesn't necessarily help polar bears. For the bears, the relationship between seal pup survival and sea ice levels means that in low sea ice years, the bears don't just lose access to their food—they lose the supply as well.

The Sierra Club hikers planned to spend two nights near Nachvak Fjord, so they'd have a full day to explore the area. On that first night, they clustered behind the electric fence in the cooking area and prepared cream of potato soup and pesto pasta for dinner. Gross and Chase's trips are known for good food, thanks to Chase's expertise as a camping chef and her arsenal of delicious, easy-to-prepare recipes.

While they worked, they watched lemmings weave in and out of tall grasses nearby. Wolves occasionally wandered into view.

Other than Isenberg and Dyer, they had all traveled with Chase and Gross before. As they cooked, the old friends told the story of how Chase had earned the nickname "Wolf Woman" on a 2005 expedition to the Yukon.

One afternoon she was walking through a meadow when she saw a pair of eyes approaching her from a pass. Another set of eyes appeared, and then another, until six sandy-colored wolves were staring down the human in front of them. For a moment Chase froze, locked in a staring contest with a seventh wolf, which had a gray coat and the biggest blue eyes she'd ever seen.

She yelled for help. But "instead of coming to her aid, we grabbed our cameras and started taking pictures!" Frankel recalled.

The wolves continued toward Chase, who raised her arms and yelled at them. They parted, ran around her, and continued on their way, leaving Chase behind, stunned by the event.

The group on that earlier trip had encountered more than a dozen grizzly bears. Once, when they stumbled upon a mother grizzly and her cub, the mother bear reared up on her hind legs. The hikers clustered together, raised their arms and started shouting. A few moments later, the bear dropped to all fours and walked away, growling, with her cub.

That night in the Torngats, dessert was blueberry cheesecake. Then the cleanup crew took over, washing their dishes in water from a nearby stream that they heated on their camping stoves.

In small groups, they left the cooking area and headed back to their tents, announcing each time the fences were turned on or off. Rodman, who is tall and lean, could step easily over the fence. For the shorter people, like Frankel, the risk of being zapped wasn't worth it, so they switched the fence off to get in or out.

That evening, Rodman called his wife back home in New York City on the satellite phone he carried whenever he traveled to remote areas. The scenery was "absolutely extraordinary," he said.

The sun sets late during the summer in the Arctic. When the sky finally darkened at about 10:30 p.m., they retired to their tents and settled down, the quiet punctuated only by the steady lapping of water onto the shore of the fjord.

At about 4 a.m., Castañeda-Mendez woke up to go pee. Trying not to disturb his wife, he slipped out of his sleeping bag and unzipped the door of their tent. When he stepped outside, he saw that he wasn't alone.

"Hey!" he called out. "Polar bear on the beach!"

A mother and her cub were walking along the shore in the early morning light. The mother bear's snout was raised in the air, sniffing out her neighbors.

Chase joined her husband while Dyer and the others grabbed their cameras. Here they were, just shouting distance from some of the world's most violent predators, yet the scene was overwhelmingly peaceful. Dyer was on the verge of tears as he watched the bears walk along the shore, the cub close on its mother's heels.

Safe in the confines of their electric fence, the hikers felt a quiet connection with animals they all knew would, in some circumstances, see them as prey. That reality, though, felt very far away. There's something about seeing an animal with its offspring that made it almost impossible not to anthropomorphize and feel like, *Here's a parent, not so different from me or my parents, taking care of its young and teaching it how to survive.* It was a powerful feeling.

They were in awe of their good fortune. They hadn't even been there a full day, and already they'd had a *National Geographic* moment.

For mother polar bears across the Arctic range, the summer of 2013 brought particular hardships.

Although polar bears don't normally hibernate, pregnant bears spend the winter in dens, rather than out on the ice, hunting. That means that in the spring and early summer before they enter the dens, they need to gain as much weight as possible.

But when the ice broke up early in 2012, the spring hunting season was cut short. Pregnant bears had less time to store up calories. When they emerged from their dens with their cubs in the spring of 2013, the last time they'd been able to hunt was in 2012—the historic low point for sea ice in the Arctic.

Studies of the western Hudson Bay polar bears have shown a direct link between the timing of sea ice break-up and the survival rates of newborn cubs. The link extends to dependent sub-adult polar bears, who nurse for about two years, until they are roughly the same size as their mothers.

"Pregnant or lactating females and their dependent offspring have the most tenuous future as global warming occurs," biologist Charles Robbins wrote in one study, published in the *Journal of Mammalogy* in 2012.

Elizabeth Peacock, a biologist working with the Government of Nunavut and the U.S. Geological Survey, flew around the Davis Strait in helicopters from 2005 to 2007, locating polar bears and darting them with tranquilizers to measure their vital statistics. The first bear she tranquilized was in the Torngat Mountains, not far from Nachvak Fjord.

In the winter of 2013, Peacock's analysis of her data, along with 35 years of capture and harvest data from the region, was published in the *Journal of Wildlife Management*.

Her first finding was, on its face, good news for polar bears in the area: population numbers in the Davis Strait were strong. With more than 2,100 bears, the Davis Strait boasted about 10 percent of the estimated 20,000 bears worldwide. That was in sharp contrast to bear populations in the southern Beaufort Sea and the western Hudson Bay, which are declining.

The density of polar bears in the Davis Strait was also far higher than in other areas with seasonal sea ice. In other regions, you could expect to find an average of 3.5 bears in an area of 386 square miles. In the Davis Strait, that number was 5.1.

One reason for the strong population numbers, Peacock hypothesized, could be the large number of harp seals in the region.

But Peacock's other findings raised questions about the population's long-term stability.

The litter size for newborn cubs, called "Cubs of the Year," was lower in the Davis Strait than in any other subpopulation. "Cub recruitment," or the survival of cubs into adulthood, was also declining. So was the general body condition of the bears.

"Declines in body condition, reproduction and recruitment are likely to precede declines in survival in a long-lived species, including polar bears," Peacock wrote.

Peacock didn't point to a direct cause for the bears' declining health. But she offered two possible causes: population density, which leads to increased competition for food, or a loss of habitat.

Essentially, Peacock found that while there's an abundance of polar bears in the Davis Strait, there isn't enough of their natural habitat—sea ice—to support them. The impact of that stark fact is reflected in the bears' reproductive systems and general health. In a year with low sea ice levels, that could add up to lots of hungry bears.

On Monday, July 22, after a breakfast of oatmeal and coffee, the hikers packed up their daypacks and headed east to explore the area around the fjord. Gross stuck one of the flare guns in his backpack. Chase carried the other.

The weather felt unpredictable, with heavy clouds settling in over the fjord and wind and rain beginning to whip through their campsite. Everyone bundled up. Frankel wore polypropylene long underwear, a fleece vest, a down sweater, a raincoat, a wind jacket, two pairs of pants, a hat and gloves.

The Torngat Mountains are technically sub-Arctic, but they lie along the 58th parallel, putting them above the tree line and within the Arctic eco-region. The group hiked through scrub willows and grassy hills and along the ledges above the campsite. The rain turned to a cool mist then gradually cleared, revealing blue skies and spectacular views of the Labrador Sea.

As usual, Castañeda-Mendez took the lead. He relished the moments alone and allowed some distance to grow between himself and the group. Gross, Rodman and Isenberg typically stayed in the middle of the pack, while Dyer, Chase and Frankel brought up the rear. They bantered while they walked. Occasionally, Gross called out to Castañeda-Mendez. *Slow down, wait up.* It was important that no one got too far from the group.

As they made their way, they found black bear scat, caribou antlers and what appeared to be a wolf skull—everyday detritus from the park's regular residents. Dyer tucked a tooth from the skull into his pocket.

After a quick lunch of beef jerky and bagels with peanut butter, honey and Nutella, they turned back. At about 3:30 p.m. they reached a wide stream near their campsite. They sat on rocks and changed to waterproof boots or left their feet bare. The water was shallow, clear and shockingly cold. For feet that had been banging around in hiking boots all day, the cool stream offered quick relief, even through rubber boots.

Castañeda-Mendez was walking barefoot, halfway across the stream, when Dyer saw something lumbering toward them.

"Polar bear!" Dyer shouted.

"Get back here! Get back here!" Chase yelled at her husband. "We have a bear!"

The animal was about 150 yards away and walking toward them. Castañeda-Mendez tromped back through the water and the group clustered together on the side of the stream, following the protocol Lagacé had rehearsed with them before they left Barnoin River Camp: *Stand together. Make yourself seem big. Make loud noises, especially metal on metal, like the banging of poles.*

The bear was larger and had a fuller coat than the female they had seen that morning. Slowly it walked toward them, nose in the air and tongue sticking out, apparently trying to assess the two-legged creatures it had stumbled upon.

Despite the group's banging and shouting, the bear kept coming. While Castañeda-Mendez fired away with his camera, Gross pulled out his flare gun.

"I'm gonna shoot," he told Chase when the bear was within 50 yards.

"I think that's a good idea," she said calmly.

The flare shot forward with a flash of light, but the bear kept advancing. It wasn't until the shell landed in front of the animal, causing a second burst, that the bear turned and took off in a dead run.

The group cheered, clapped and banged their poles together, celebrating their victory. It was like scoring a touchdown at a football game.

But the bear didn't go far. It settled on a ledge about 300 yards from their camp and lay there quietly, watching them.

By the time they reached the safety of their fence, the rain was coming down hard. Most of the group settled into their tents for a nap before dinner. But Dyer was uneasy. He couldn't relax with the bear so close.

Dyer stationed himself outside his tent, leaning on his poles and staring down the bear as it watched them. Castañeda-Mendez said he looked like one of the guards at Buckingham Palace. Dyer stood watching the bear for more than an hour, drenched under the dreary gray sky as the afternoon waned.

Finally, the bear and the rain wore him down. Gross and Isenberg were watching the bear from inside their tents, so Dyer retired to his tent, too. He opened "Leaves of Grass"—the only book he'd brought with him—but soon fell asleep.

5

The cold, rainy afternoon became a cold, rainy evening and still the bear watched them.

After his nap, Dyer walked a few steps through the drizzle to the tent next to his, where Chase and Castañeda-Mendez were relaxing. He had just read a poem that felt so right he had to share it with someone. Through the nylon walls of the tent, he read to them Whitman's "Me Imperturbe."

Me, wherever my life is lived, O to be self-balanced for contingencies!

O to confront night, storms, hunger, ridicule, accidents, rebuffs, as the trees and animals do.

At about 5 p.m. the campers made their way across a rocky strip to the cooking area.

Up on the ledge, the bear appeared to be lounging. Using the zoom lenses on their cameras, they watched it roll on its back and then lie on its belly, resting its head on its crossed forelegs. To Frankel, it looked like a big dog. To others, including Dyer, it looked like a menace.

The hikers couldn't help but compare the polar bear that was watching them with the grizzly bears they were familiar with. Both species seemed curious about humans but not drawn to them. Both responded to loud noises or flares.

These reasonable conclusions, however, didn't take into account what makes a polar bear a polar bear. Although polar bears and grizzlies share the vast majority of their DNA, it's the differences between them, the subtle and not-so-subtle ways in which they've diverged, that define the polar bear.

Biologist Andrew Derocher began studying polar bears in 1984, when his work on grizzly bears led him to their Arctic descendants. In his 2012 book "Polar Bears: A Complete Guide to their Biology and Behavior," he imagines a scenario in which, sometime within the last 6 million years, grizzlies wandered onto shore-fast sea ice, either because they had extended their habitat into areas with cooler climates or because a cooling period had occurred. On the ice, they would have come face to face with seals, which at that point had no known predators and no reason to fear them. It would have been open season on this fat-filled, ample prey, but only for bears that were predisposed to adapt to the ice.

Gradually, grizzly bears evolved into polar bears.

"Presumably," writes Derocher, "lighter colored bears fared better at catching seals until whiter bears became the norm."

Polar bears' hair isn't actually white, it's translucent, which allows it to radiate reflected light. The hairs are hollow—a trait polar bears share with deer—which helps insulate them and adds buoyancy when they're in water. Beneath that hair, the bears' skin evolved from pink to black, although scientists don't know exactly why.

The bears' hind feet sprouted an abundance of fur to provide traction and warmth. Their paws became webbed and their leg bones became denser than grizzlies', allowing them to swim long distances and helping them earn their scientific name, *Ursus maritimus.*

The bears' skulls changed, too. They became longer and narrower, allowing the air the bears inhaled to be better warmed and helping them capture prey in tight spots, like birth lairs and breathing holes. Their eyes became slightly elevated, a possible adaptation for the aquatic lifestyle.

The Arctic climate made it harder for females to carry large litters to term, so the litter size shrank, from three to four for grizzlies to an average of two for polar bears. As the litter size shrank, so did the number of teats for nursing. Grizzlies and American black bears have six; polar bears have four.

Polar bears grew larger, too. Although some sub-populations of grizzlies are as big as large polar bears, polar bears are generally larger than grizzlies or any other bears—and are among the world's largest mammals.

These physical changes happened in tandem with changes in the bears' habits, Derocher writes in his book.

Since life on sea ice didn't expose the bears to the flora that grizzly bears feed on, polar bears became carnivores, with a hunting style tailored for routing sea creatures from beneath the ice. Their skulls—now too narrow to allow them to effectively chew coarse vegetation—were ideal for this work.

Hunting was best in cold weather, which gave them access to their prey through the thick layer of sea ice. The bears stopped hibernating in the winter, except for expectant mothers.

These changes happened over millennia—but just how many millennia remains an open question. The oldest polar bear fossil, a jawbone, is between 110,000 and 130,000 years old. Analysis of the jawbone showed that this ancient bear's diet was similar to that of modern polar bears. Its primary prey was seals and small whales, rather than the mix of fresh water fish, land mammals and plants that brown bears eat.

While we don't know exactly how long it took for natural selection to skate its way across the sea ice and form a species uniquely prepared to thrive there, the science is clear on what's happening now. The sea ice has up and changed on the bears, in the course of just a few decades. The question now, of course, is how will the bears respond?

Half of the Sierra Club hikers helped prep and cook dinner that night: cream of potato soup, pesto pasta, cheesecake with blueberries and almonds. The others handled cleanup.

Remote locations and small groups make for fast friendships, and they were already comfortable enough with each other to mimic the soft drawl of Dyer's Maine accent and share stories of past trips and their lives back home.

They didn't talk much about the bear that was still watching them from the ledge. It seemed almost like a piece of the landscape—just one more detail in their majestic setting.

Castañeda-Mendez felt reassured by the bear interactions they'd had that day. The mother and cub hadn't seemed the least bit interested in them. And the bear up the ledge had responded to the flare in the way they had hoped it would.

But Dyer couldn't shake his sense of uneasiness.

"Why don't we post a watch?" he asked Gross after dinner. They could take two-hour shifts overnight until the bear was gone, he suggested.

But Gross wasn't worried. "That's what the fence is for," he told Dyer. They had checked the fence, to make sure the wires were taut and the battery was switched on. Gross and Chase had done their homework; they were sure the fence would keep them safe.

Dyer thought back to their orientation at Barnoin River Camp where he remembered Lagacé telling them, "Polar bear touches that, you won't have to worry."

The group tucked into their sleeping bags. Just the thin nylon of their tents separated them from the vastness around them and the polar bear watching from 300 yards away.

They had faith in their fence.

Rodman woke in the middle of the night to the sound of screaming. For a moment he was frightened. Then he realized it was Chase, in the tent next door, having a nightmare. He could hear Castañeda-Mendez soothing her.

Isenberg slept fitfully, too. Each time he woke up, he checked to see if the bear was still there.

"Sure enough he was, sure enough he was…" And then, about 1 a.m., "He wasn't." It was unnerving, not knowing where the bear was, but there was nothing Isenberg could do about it. He went back to sleep.

The next morning was cold and rainy. The hikers bundled into layers—jackets, waterproof pants, winter hats and gloves—and hung around the campsite for a couple of hours, hoping the sky would clear so they could move on to their next location. When it became evident the weather wasn't going to cooperate, they packed up their daypacks and went exploring again.

This time they hiked northwest, in the direction they would be heading the next day. From one high point, they looked down on the valley where they planned to camp.

Again, they were surrounded by wildlife: whales in the fjord, caribou, ptarmigan, black bear scat full of berries. By afternoon the weather began improving and they peeled off some of their warm layers. They stopped at a rock perched high above their campsite to take silly pictures of each other.

When they got back to camp, Chase unpacked a little happy hour for them to enjoy: salami, crackers and Bacardi 151 rum mixed with lemonade. "It was a celebration," she said. A celebration of being in a beautiful place, far from civilization, with people whose company they enjoyed.

As they cooked dinner that night, they talked excitedly about moving on the next day. Before Gross turned in, he made sure everyone was set for the night and walked the camp's perimeter, confirming that the electric fence was still working.

Isenberg detached the fly from his tent so he could watch the clearing sky. It was too early in the year for the Northern Lights to appear, but he wanted to be ready, just in case.

Dyer pulled on his silk long underwear and slid into his sleeping bag. He'd always been able to fall asleep anywhere, and this night was no exception.

Before Gross crawled into his sleeping bag, he tucked his flare gun into his boot, as he did every night. He fell asleep listening to the waves against the gravelly beach of Nachvak Fjord.

6

At 3:30 a.m. the campers woke to screams of pure terror, screams that came all the way from the gut.

"Help me! Help me!"

Gross's hands flew to unzip his sleeping bag and he grabbed the flare gun from the boot near his head. He had no idea what was going on, but he knew it wasn't good.

From the little window of her tent Chase saw the shape of a polar bear just a few feet away, standing over the tent beside her. It was down on all fours, eye-level with Chase, huge and white except for the black of its eyes and nose. It turned and stared right at her.

"RICH!" she screamed.

Chase grabbed her flare gun while Castañeda-Mendez raced out of the tent. The bear was just a few yards away, biting at the tent next door and then dragging it into the darkness.

Gross ran into the grass in his long underwear and aimed the flare gun toward the bear as it started running. It was a moving target, now 75 feet down the beach, heading west, parallel to the shore of the fjord. Something was dangling from its mouth.

Isenberg shot out of his tent. Rodman scrambled for his glasses and rushed outside, too, still in his underwear and bare feet. Frankel fought frantically with the zipper of her tent fly. Finally, she ripped it open and ran out in her long underwear.

They could see only a few yards away. But they saw enough to know that the thing in the bear's mouth wasn't a thing at all. It was one of them—it was Dyer.

Gross aimed the flare gun toward the bear and fired. The flare erupted and then a second brilliant bright light exploded in front of the bear. It dropped Dyer and took off running.

But the bear didn't run far. After another 75 feet or so, it stopped and turned around. Dyer lay crumpled on the ground. It was coming back for its prize.

Gross reloaded and fired again. This time, the bear ran off into the distance.

Nachvak Fjord echoed with the group's screams and cries. They were one part fear, one part desperate attempt to keep the bear from coming back.

They had to get to Dyer. But the light was bad and there was no telling whether the bear might come back. Rodman remembers seeing the fence sparking on the ground.

Frankel screamed Dyer's name, but he didn't respond.

7

"**G**o away bear, go away bear. Just go away."

Dyer silently repeated that mantra as he lay on the gravel, listening to the footsteps of the bear nearby.

He wanted to let his friends know where he was—to yell to them, to wave. But his jaw wasn't working and he could barely move his arms.

The attack had happened so fast. He had been sound asleep when something—he wasn't sure what—caused him to stir. As his eyes adjusted and he looked up at the top of his tent, he saw two large arms, silhouetted by the bright Arctic moon, sweep across the thin nylon. Polar bear paws can be a foot wide, with claws up to two inches long. There was no mistaking what was happening.

Before Dyer could react, his tent was lifted off the ground, with him in it. He heard the fabric tearing.

"Bear in the camp!" Dyer remembers shouting. "He's got me! Oh, he's got me!"

The bear ripped him out of the tent, the animal's mouth clamped around the crown of Dyer's head. Dyer heard his jaw break and felt the bear's teeth, which can grow to an inch and a half long, puncture his head and neck. He could smell the fishy, oily stench of the animal's saliva.

The bear dragged him toward the mouth of the fjord, Dyer's face bumping against its chest. He stared at its white stomach and the yellow stains on its hindquarters as it carried him farther and farther from camp. He noticed with detachment that one of his socks had fallen off.

And then there was noise coming from behind him—the shouts of his friends.

The bear turned its head toward the sounds, flipping Dyer into the air and slamming him against the ground. Without losing its grip on Dyer's head, it continued moving toward the water.

Dyer descended into himself, overcome by a sense of calm.

Well, you're gonna die now, he thought. *Everyone is going to meet this time ... I wonder what it's going to be like.*

He had sometimes thought about how it would feel, that terrifying last moment before you die. But instead of fear or panic, he found himself filled with an overwhelming sense of calm, a complete and utter acceptance of his fate. In the arms of something so strong, there was no point in holding on to hope.

At that moment, his head still in the polar bear's jaws, Dyer saw a flash of light and heard the unmistakable *whoosh* of a flare gun. The bear dropped him—hard.

He heard vertebrae in his neck crack as they broke. He was in shock, mercifully, and couldn't feel the pain.

Some people pray in life-threatening moments, but Dyer wasn't religious. As he heard the bear walking away, he simply pleaded: *Go away bear. Just go away.*

The sounds of the animal's massive paws on the gravel began to fade. Then they stopped and grew louder again. The bear was coming back.

A second flare fired—*whoosh*—lighting up the sky. The bear took off, its plodding footsteps growing faint in the distance.

Dyer lay motionless on the ground, trying not to make any sound or movement that might draw the bear back to him.

8

From behind the broken fence, Chase drew on her emergency response training and urged the group to assess everything they could see. "Where's the bear?" she yelled as they all stared into the darkness.

Gross handed his flare gun and binoculars to Frankel so she could cover him. "I've got to get out there," he said. Isenberg went with him.

As the beams of their flashlights swept the ground, they saw pieces of Dyer's tent. His clothes. His sleeping bag.

About 75 feet from the campsite they found the lawyer's crumpled, blood-drenched body. At first, they thought Dyer was dead. But when Isenberg knelt beside him, he saw that Dyer was breathing.

They tried to lift him, but he was too heavy. They called for help. Castañeda-Mendez and Rodman, now dressed, ran out to them. The four men laced their arms under Dyer's body and carried him back to camp. Dyer hung limply between them, a dead weight.

Castañeda-Mendez grabbed a sleeping pad from his tent, and they carefully laid Dyer down on it in the middle of the campsite.

Castañeda-Mendez covered Dyer with the double sleeping bag he and Chase shared and placed a duffle bag under Dyer's head for a pillow.

Gross and Castañeda-Mendez ran to the cook tent and pulled up its stakes. The teepee-style shape of the tent gave Isenberg enough room to work and protected Dyer from the wind. Although Chase had wilderness first-responder training and certification, Isenberg was a physician, and she deferred to him.

All he had to work with was a basic medical kit with typical first-responder materials—four-by-four gauze pads, a roll of gauze strip, antibiotic ointment, splints.

"Can you move your hands?" Isenberg asked Dyer. "What about your feet?"

He could. Isenberg cut off Dyer's shirt. It was wet with blood, and the doctor was afraid he wouldn't be able to keep his patient warm. Dyer's face was swollen and bruised, and his jaw was displaced. The good news was that he was talking.

"Thank you. Oh, thank you," he said over and over, his voice a raspy whisper because of his crushed jaw.

Much of the blood appeared to be coming from his head and neck, but the wounds were hidden beneath his long ponytail. Isenberg hacked through the blood-soaked hair with a small pair of first aid scissors.

That done, Isenberg assessed Dyer's wounds. Puncture wounds ringed his face and head, but they were oozing blood, rather than pumping it—a good sign. His arms and hands were covered in gashes. Isenberg could only assume that Dyer had been fighting back.

The biggest wound was a gash on Dyer's neck that looked as if it had been filleted open by the bear. As Isenberg examined the wound, he could see Dyer's carotid artery, the principal blood supplier to the head and neck. The artery was intact, but if anything caused it to tear, there would be nothing he could do to keep Dyer from bleeding to death.

Isenberg moved and spoke with confidence, but inside he was terrified. Not only was Dyer in critical condition; not only were they in an isolated area several hundred miles from any kind of help; but Isenberg hadn't practiced medicine in 15 years.

The doctor fell into a routine—find a wound, clean it, bandage, move on. But he also realized that Dyer's wounds might not all be external. The lawyer was struggling to breathe and regularly spitting up blood and mucus. If he stopped breathing, Isenberg would have to perform a tracheotomy, punching a hole in Dyer's windpipe to open up the airway. But how?

Isenberg took a mental inventory of the supplies they had on hand. Cutting the hole open would be easy—he had a pocketknife that would work for that. But what would keep the hole open? A drinking straw wouldn't work; its diameter is too small to allow enough air in. He cycled through other possibilities until he hit on something: the drinking tube in a Camelback. The tubes to the water pouches are sturdy, and definitely big enough.

The prospect of actually performing the procedure was terrifying, but at least he had a plan. He prayed that he wouldn't have to use it.

Isenberg quickly used up the supplies in the medical kit. There was little more Isenberg could do there in the wilderness, without sophisticated medical equipment.

Isenberg held Dyer's hand and prayed as his patient slipped in and out of sleep.

Outside the tent, Chase forced herself to stay calm as she worked her satellite phone, using the list of emergency numbers and protocols she'd been given by Parks Canada. Rodman helped her.

Chase's biggest fear was that no one knew exactly where they were. With the whipping winds and cloud cover returning, getting a rescue plane in would be difficult.

Her first call was to the number at the top the list—the closest outpost of the Royal Canadian Mounted Police in the village of Nain. It was about 200 miles away as the crow flies, but separated by the Torngat Mountains.

At 3:45 a.m. Atlantic Daylight, Chase got a police dispatcher on the line. Her group had been attacked by a polar bear, she said. One member of their party needed to be evacuated, and the rest of them were in danger.

Their electric fence—which clearly hadn't worked—was in tatters. The fence around the cooking area was still intact, but moving there seemed pointless, given that they no longer believed the few strands of wire could protect them.

They needed more flare shells, she said. They'd started off with eight shells, but four bears and one attack later, only five shells remained. They still had bear bangers and spray. But after watching the bear drag Dyer out of camp like a ragdoll, they didn't think they'd help much.

While Chase stayed on the phone, Frankel made slow circles around the perimeter of the camp, holding Gross' flare gun, her eyes scanning the horizon. Castañeda-Mendez and Rodman took turns patrolling with the second flare gun. They also cooked food, heated water and collected what remained of Dyer's gear.

Gross stationed himself just outside the cook tent, ready to get whatever Isenberg needed, and made rounds within the group, ensuring that everyone felt safe. Every 15 minutes, Chase called the police dispatcher to update their status and ask about the rescue plan.

Three thoughts cycled through everyone's minds.

They needed to keep Dyer alive.

They needed a rescue plane as soon as possible.

They needed to be vigilant in case the bear came back.

9

At 4:20 a.m., the unspeakable, unshakable fear that had been driving them for the last hour finally eased. Dyer was stable, Isenberg announced. If his carotid artery didn't rupture and he kept breathing—and if the rescue team arrived relatively soon—he would survive.

The sun was coming up. If a bear came their way now, at least they'd be able to see it.

But the rescue operation wasn't coming together smoothly.

Chase called the police at regular intervals, keeping a log of every call. But a disturbing pattern had emerged. Often, someone new answered the phone, someone totally unfamiliar not only with their circumstances but with basic information like the location of Torngat Mountains National Park, and the fact that a float plane or helicopter was needed to evacuate Dyer.

At 5:30 a.m., an hour and 45 minutes after her first call, Chase gave a dispatcher the phone numbers for Base Camp—the outpost that serves as Parks Canada's base in the Torngats—and for Alain Lagacé at the Barnoin River Camp.

At 6 a.m., she suggested to another dispatcher that a helicopter be sent from Base Camp for Dyer and that more flares, or a person with a gun, be brought in to guard the group.

A half hour later, someone told her that was what they were trying to arrange.

Although Chase didn't know it, Base Camp had been alerted at 6 a.m. and was already working on plans, backup plans and backup-backup plans.

Gary Baike, a Parks Canada employee, had awakened a medic, Larry Brandridge, from his tent and told him to get ready to go. Baike and Base Camp manager Wayne Broomfield checked the hiking route the group had filed with Parks Canada and figured out where they might be. When they learned through a dispatcher that a doctor was traveling with the group, the level of intensity ratcheted down "from an emergency to an urgency," Brandridge said later.

Base Camp consists of a couple of year-round buildings and a few dozen tents for sleeping. Everything sits behind a 5-foot-tall electric bear fence that is connected to an alarm system. Because the camp is surrounded by mountains, the only way in is by boat or helicopter.

On clear days, a helicopter makes regular runs from the camp, taking visitors and park staff to various islands and locations in the park. But everything stops when the weather is bad, as it was that day. The sun hadn't come up over the hills yet, and everything above 500 feet was completely fogged in. There was no way they could get a helicopter up.

As a backup, they started preparing a speedboat to take Brandridge and Jacko Merkuratsuk, a Parks Canada employee who is also a bear guard, up to Nachvak. The boat trip would be much longer than the helicopter ride, but if the weather stuck around all day, it might be the best option.

At the same time, they dispatched *The Robert Bradford*, a commercial fishing boat they contracted for the summer, to make the 10-hour trip up to the fjord. Unlike the helicopter *The Robert Bradford* was big enough to get all of the hikers out together.

Back at the fjord, Dyer woke up occasionally, murmuring thank-you to whoever was with him at the time. Isenberg occasionally barked orders—asking for a towel, or water or a bottle for Dyer to urinate in. Moments later, a hand would reach under the tent, holding whatever was needed.

The others kept watch and stayed busy heating water for tea or coffee and preparing food for anyone who was hungry. They took turns carrying the flare guns and keeping watch.

By 7:30 a.m., the clouds were lifting. But a police dispatcher told Chase that the fog remained thick over Base Camp and the helicopter there was still socked in.

At 7:45 a.m. Chase finally reached Lagacé at Barnoin Camp. If Base Camp couldn't get to them, maybe Lagacé could. But the fog was thick there too, he said. There was no way he could fly.

At 8:10 a.m., Chase was shocked to hear yet another new voice on the police line. The man had no idea who Chase was, why she was calling or what had happened.

The new dispatcher explained that Chase's earlier calls had been forwarded to the police department's after-business-hours line about

950 miles away in St. John's, Newfoundland, where the staff was unfamiliar with the Torngats and unaware of just how remote their location was. Now that it was after 8 a.m., Chase was talking with the police dispatcher in Nain, just 200 miles away and the natural place to respond.

It had been nearly five hours since the bear had pulled Dyer from his tent. For five hours they'd been listening to his raspy voice and labored breathing. Five hours without a rescue. And they were starting over now?

Rather than tell her story again, Chase tracked down the last dispatcher she'd worked with in St. John's. When she reached him, he had good news: The sky over Base Camp had cleared and a helicopter was on its way with a medic and a bear guard on board.

Minutes later, the group heard the *thump-thump-thump* of the chopper's blades and saw it moving through the mountains and across the fjord toward them.

From inside the tent, Dyer heard someone shout, "Here comes the helicopter!"

Outside, the others waved their arms and jumped up and down, signaling for it to land.

Brandridge, a former cop with broad shoulders and a stocky frame, got off the helicopter and huddled with Isenberg to get an update on Dyer's status.

Brandridge ducked into the cook tent and found Dyer conscious and checkered with bloody bandages. He had a laceration over one eye and the lower part of his left earlobe was gone. But all things considered, Brandridge was pleasantly surprised. "I was expecting chunks of meat missing, more puncture wounds," he said later.

When Brandridge told Dyer that they were going to fly him out, Dyer's eyes brightened. "Good!" he rasped.

Brandridge maneuvered Dyer onto a scoop stretcher—a clam shell-like contraption that can open up and be placed around a patient—and with the other men put the stretcher on a backboard. Chase held onto Dyer's hand as they carried him to the helicopter.

As the helicopter rose through the clouds, the people they left behind became tiny specks and then disappeared entirely.

Isenberg looked out the window, counting polar bears and black bears along the way.

Back on the ground, Merkuratsuk began gathering wood and building a fire, his gun slung over his shoulder. The level of fear in the camp dropped a few notches. A fishing boat was on its way from Base Camp. If the weather cooperated, they should be safe on board by late afternoon.

The idea of spending another night on the fjord was too terrifying to contemplate, particularly when Merkuratsuk told them what he had seen before the helicopter landed: a large polar bear walking in the area where the group had planned to hike that day.

Jacko Merkuratsuk knew Nachvak Fjord—and polar bears—as well as anyone in the region. When he and his nine siblings were growing up in Nain, they spent most of their summers at the fjord. One of his brothers had been born there.

Three of his siblings, as well as his son, were bear guards, licensed by Parks Canada to travel into the park as guides.

It was Jacko Merkuratsuk and other local Inuits who had helped inspire the population study of the Davis Strait polar bears that biologist Elizabeth Peacock had launched in 2005. They seemed to be seeing more polar bears in the region, and if they were right, they wanted Parks Canada to raise the annual quota for polar bear hunting. At that point, the five Inuit communities in Nunatsiavut could kill a total of six bears a year, a quota that had been in place since 2001.

The hunting licenses are distributed via a first-come, first-served system to members of Nunatsiavut's native communities. Hunters are given 72 hours to try to kill a bear before they have to return the license so it can be passed along to someone else. Once the quota is met—six bears in the past, or 12 bears now in 2014—the season ends. Different quotas are set for the other regions of Canada where polar bears are hunted.

Though this system is controversial outside native communities, it allows for the continuation of a traditional way of life that assigns special cultural capital to polar bears. Killing a bear can also result in a financial windfall. Good-paying jobs can be hard to come by in these remote areas, and raw hides go for $8,000-$10,000 on international markets, and have sold for as much as $22,000. The meat of the bear is eaten, too.

Elders in native communities are a valuable resource for researchers trying to understand the implications of climate change for the region and for the bears.

In 2010 Moshi Kotierk, a social science researcher with the Department of Environment in the government of Nunavut, questioned 31 Davis Strait hunters and elders about polar bears and climate change.

Twenty-four reported seeing more bears and agreed with the statements: "There are problem polar bears now." "To sleep in tents is concerning. I won't sleep in tents any longer." "They are in or around communities." Eleven didn't remember seeing polar bears in the Davis Strait in the 1940s, '50s or '60s.

Few connected the increase in bear-sightings with climate change, but the majority agreed that the sea ice has changed significantly.

Twenty reported that the ice doesn't form as well as it used to, agreeing with the statement that "ice that we use[d] to travel over, we can't." Eight believed the loss of ice had led to an increase in the number of bears.

In Labrador, unpredictable ice can have deadly repercussions. Residents of Nain and other communities around the Torngats rely on snowmobiles—they call them Ski-Doos, after the company that makes them—to get around in the winter. The Ski-Doos are their cars; the ice is their highway.

But the ice that once provided sturdy, safe passage is no longer fully trustworthy. A key element of climate change is variability—not only is the climate changing on a grand scale, over years and decades, but it's also changing on an immediate scale, with higher highs and lower lows. In addition to freezing later and melting earlier, the ice now varies during the season.

In Nain, locals say a Ski-Doo can be flying across ice that appears normal when suddenly the ice cracks and the sled crashes through to the frigid ocean below.

Although the Merkuratsuks rarely mention climate change directly, their familiarity with the ecosystem allows them to identify the shifting line that separates safe and dangerous travel in the Torngats. Particularly when it comes to polar bears.

While Gross and Chase knew from their research that bear guards carry guns to protect groups from attacks, they didn't know that bear guards can also determine whether a bear has been in the area recently. The Merkuratsuks know which areas along the fjord polar bears are drawn to, and which should be avoided. They know not to walk in the willows—an area that some of the Sierra Club hikers used for their bathroom—because polar bears go there to keep cool from the summer sun.

They also know the routes that polar bears walk, including the route that Chase and Gross's group was planning to take—a path called a "polar bear highway" by one Parks Canada official.

And when the Merkuratsuks do encounter a bear—say, one that's been lounging on a nearby ledge for hours—they recognize it as a sign of danger. They know it's time to pack up camp and move on.

But Chase and Gross didn't know what the bear guards knew. None of their research, including their talks with Parks Canada, indicated just how valuable a bear guard could be. And in this case, what they didn't know nearly killed them.

The helicopter touched down on the landing pad at Base Camp around 8:30 a.m. As the propellers slowed to a halt, an ATV with a trailer pulled up. Dyer's stretcher was loaded onto a mattress in the trailer and driven to the medic tent.

With Isenberg at his side, Brandridge inventoried and cleaned Dyer's wounds. He began with the bite and claw marks on his face, which were dripping blood into Dyer's eyes. After removing each bandage, Brandridge cleaned away the blood and photographed the wound.

"How are you feeling?" Brandridge asked.

"Like shit," Dyer said.

"That's not bad for someone who just got attacked by a polar bear."

After finishing with Dyer's face, Brandridge peeled off the bandage covering the big wound on his neck.

The thick odor of meat immediately filled the tent—an odor that, to Brandridge, smelled like death.

He saw that the hole in Dyer's neck was about the width of a pencil and went behind his jugular and toward his esophagus. Each time Dyer inhaled, he was wicking blood into the wound.

Trying to keep his voice calm, Brandridge asked Isenberg to watch Dyer. Then the medic rushed to the main office of Base Camp.

The plan to wait for a medevac plane from Goose Bay, Labrador, wasn't going to work, Brandridge told Baike and Broomfield. Based on his assessment, Dyer's condition wasn't stable after all. His lungs could be filling with blood as they spoke.

Brandridge could clean him up and make him comfortable, but he didn't have the medical equipment or expertise for the kind of operation Dyer needed to save his life.

After considering their limited options, they decided to put Dyer back on the helicopter and send him to George River, a town about 45 minutes away, where a first-response team with more sophisticated equipment would meet them. From there he'd be flown to Kuujjuaq and then on to Montreal.

Within about a half hour, the helicopter was up in the air, weaving through the passes between Base Camp and the river valley en route to George River. The weather was relentless, with snow in one pass, and sleet and rain in others.

Brandridge radioed Base Camp to say they were making progress. But he couldn't help but feel that it was unlikely Dyer would survive.

As the helicopter reached the Korrack River Valley, the weather improved. They followed the river to George River, where an ambulance met them at an airstrip.

At about 4:40 p.m., 13 hours after the polar bear had attacked the Sierra Club hikers, *The Robert Bradford*, a vision in green and white, dropped anchor in Nachvak Fjord.

The fishing boat's owners, brothers Chesley and Joe Webb, welcomed the five remaining hikers and Jacko Merkuratsuk aboard. Beside them was Cappuccino—called Chino—a fluffy white dog that looked as if it belonged in a suburban home, not on a Labrador fishing boat.

The fog was thick and the rain was cold. Most of the group huddled in the cabin, at a small kitchen table with bench seats. At the front of the cabin Chesley Webb sat at the captain's wheel and steered the boat through the nasty weather and toward the mouth of the fjord. Frankel stood beside him, looking straight ahead and trying not to succumb to seasickness.

Joe Webb wore thigh-high black waders and a long green rain jacket as he worked on the deck. Two long lines with anchors hanging off of them—called "birds"—dangled from the sides, helping steady the boat in the choppy water. Still, the boat rocked, turning the stomachs of the group and making them generally miserable.

At 7:30 p.m., they neared the mouth of the fjord. But instead of entering the open waters of the Labrador Sea, the Webbs dropped anchor. With little to no visibility and icebergs dotting the coastline, it wasn't safe to continue in the dark.

Though the Webbs no longer use their boat for commercial fishing, the white-haired, blue-eyed brothers are born fishermen. They eat what they catch or hunt—black duck, ring seal, fowl, Arctic char. That night, freshly caught Arctic char was on the menu, "the most delicious char you'd ever have," Rodman said.

They tucked in where they could for the night. Frankel and Chase slept in the small, angled bunks below deck. Rodman, Castañeda-Mendez and Gross lay on the floor of the galley, fitting themselves into the small space carefully, like puzzle pieces. Eventually, Castañeda-Mendez was so uncomfortable that he gave up and moved to the galley table. He sat on the bench and laid his head on the table. Gross joined him there later.

They covered themselves with whatever sleeping bags they could find, as wind from the deck found its way through the door into the cabin.

Toward morning the Webbs pulled up anchor and *The Robert Bradford* steamed into the Labrador Sea. At 5 p.m. after a long day of high waves and bad weather they pulled into Saglek Bay with Base Camp in sight.

That night, their most basic needs were met: A warm meal. Showers. Cabin-like tents to sleep in. All behind the formidable bear fence that surrounded the camp. They felt safe.

The next day, while they waited for the plane that would take them to Kuujjuaq, they set off for one last hike, this time flanked by two bear guards. The climb up Base Camp Mountain was a short but steep ascent through snowy passes and into clear, bright skies. At the top, one by one they used rocks to form a makeshift Inukshuk, a statue of stones traditionally used in native Arctic communities to communicate something—an indicator of safe passage, the location of a path or a marker for a reserve of food.

Inukshuks can also have a spiritual purpose, sometimes used as a monument or memorial to an event or person.

The wind blew back the brims of their hats as they piled their rocks high. Below them, they could see Base Camp safe within its fence. Just to the north, they could see the snow-crested mountains of Torngat Mountains National Park.

Decades of studies about polar bears and the rapidly changing climate have led to a prevailing scientific narrative about the bears' future: the loss of sea ice, driven by man-made climate change, will eventually force them ashore for such long periods of time that the species is inevitably doomed.

In recent years, however, a competing narrative has emerged, driven in large part by the work of Robert "Rocky" Rockwell, a biologist and ecologist at the American Museum of Natural History in New York.

Rockwell has studied various species in the lowlands of the western Hudson Bay for 46 years, and in 2009 he began publishing his findings about polar bears and their eating habits. He suggests that because they bears are "opportunistic eaters," they might be able to survive climate change by foraging for food during their extended periods on land. His work has captured headlines in major media, because it runs counter to previous reports about polar bears. It has also been adopted by climate change skeptics as proof that the bears aren't threatened by climate change, although Rockwell himself doesn't draw that conclusion.

Rockwell writes that the western Hudson Bay bears don't necessarily fast during their months off the ice, when their hunting skills are thought to be almost useless. Some gorge on snow goose eggs and even the geese themselves. They also eat berries and plants.

Rockwell points out that while a polar bear can't run down a caribou, he has seen bears wait for a herd to pass by and pounce on stragglers. He also has seen bears stalk sleeping seals on land. Locals in the Torngats report similar incidents, as well as occasional sightings of polar bears catching char in streams, much as their grizzly ancestors fished for salmon.

"I find it a little bit crazy that people say 'No, no, the polar bears only eat seals on the ice and when the ice goes the polar bears will have less to eat,'" Rockwell said in an interview. "Well, they'll have less seal to eat, but they'll opportunistically eat other stuff."

Rockwell's critics say polar bears have long been known to find alternative food sources while they're ashore. They also say that the group's Hudson Bay studies focus on such small sample sizes—in a few cases just 10 bears—that the data can't be extrapolated to predict how polar bears are responding to climate change globally.

Perhaps the biggest flaw they see in his work is the idea that terrestrial eating—eating on the land while off the ice—is helping bears in the western Hudson Bay. In addition to having decreased body condition and lower reproduction rates, the bear population there has declined by 22 percent since the early 1980s as the ice has broken up earlier.

"If terrestrial feeding was the savior for polar bears, why are polar bears starving on land during the ice-free period?" asked Derocher, the biologist who has studied polar bears for more than 30 years.

One way to understand what may happen to bears in the future, Derocher says, is to understand what happened to them in the past.

Ten thousand years ago, polar bears lived in the Baltic Sea, around Sweden, Denmark and near Finland. As the climate warmed, there wasn't enough ice there to sustain their population, so they followed the ice north.

"They didn't hang around to try to exploit terrestrial resources," Derocher says. "This is a sea ice-obligate species and once the sea ice dropped below that threshold, that just didn't give them enough access to prey."

The big question, of course, is whether polar bears can adapt to the latest changes in their habitat.

Brendan Kelly, former deputy director for Arctic science at the National Science Foundation, looks at this question in the context of the pace of change. A species' ability to evolve in the face of new conditions depends in large part on the length of its life span, he explains. If the habitat changes slowly and multiple generations can survive during that period, traits that prepare the species to thrive in the new habitat will flourish while other traits gradually disappear.

But when a species with a long life span, like polar bears, is confronted with a rapidly changing environment like the melting of the Arctic, adaptation is less likely.

Polar bears in the wild live an average of 15 to 18 years, with some living into their 30s. That means that a 15-year-old bear, living right now on the Davis Strait, has already seen its ice-free season extended by approximately two weeks. The spring melt comes about a week earlier than when the bear was born and the freeze-up comes about a week later.

"When environments change really abruptly, more typical than adaptation, which takes just incredible luck, is extinction," Kelly says. "I'm in the camp that would argue that we're really flirting with danger here because we are changing the sea ice environment. We're radically changing it, in just a few generations of bears."

The loss of sea ice, and its implications for polar bears and other large mammals that depend on the ice, Kelly says, is less like the evolution of plants over millennia and more like the abrupt environmental changes caused by the arrival of a meteor.

"We can't tell you what the outcome will be," he said. "All we can say is the prudent decision would be to try to slow this down."

10

In George River, Dyer was loaded into an ambulance and taken to the town's infirmary, where nurses got an IV into him, gave him oxygen and put a stabilizing collar around his neck. The clinic had no doctor, but Isenberg said an x-ray technician who visited every two weeks happened to be working. She x-rayed Dyer's chest and gave them a preliminary report: One of Dyer's lungs appeared to be damaged.

Before the nurses could do much else, a plane landed with a medical team from Kuujjuaq, the Inuit community the Sierra Club hikers had flown through on their way into the Torngats. For Isenberg, seeing the medical team come in was like seeing "the cavalry coming over the hill." They quickly loaded Dyer onto the plane and headed for Kuujjuaq.

Doctors at the town's small hospital discovered that Dyer's lung was, in fact, punctured.

While the team worked on him, Dyer repeatedly thanked them for their help. He pointed across the room at Isenberg. "That man saved my life," he said.

Dyer was put into a medically induced coma, and a breathing tube was inserted into his throat. At about 8 p.m., the Challenger, Quebec's flying intensive care unit, arrived to take him south, to Montreal.

Isenberg stayed behind to wait for the rest of the group. As he walked around the village he tried to process all that had happened in the last 16 hours. Wherever he went, people stopped him to talk about what had happened. Bear attacks are rare, and news travels fast in small, isolated communities like Kuujjuaq.

Around midnight on July 25, about 20 hours after being attacked by the polar bear, Matt Dyer was admitted into intensive care at Montreal General, still in a medically induced coma.

He had two broken vertebrae in his spine, but they were in his neck, high enough that the doctors weren't worried about paralysis. His jaw was crushed. His left hand was broken in several places. His right lung had collapsed. He had at least a dozen puncture wounds, including the gaping hole in his neck. A tendon in his right arm was punctured. Two arteries in his brain were occluded—permanently clogged—but his remaining arteries had taken over and his blood was flowing fine.

At 12:30 a.m., Dyer blinked his eyes open as he was brought out of the medically-induced coma. He focused under the fluorescent lights, and the first thing he recognized was the face staring back at him: his partner of 25 years, Jeanne Wells. Seeing her meant one thing: He was safe. It was over.

Dyer couldn't speak with the breathing tube in his throat. But the medical staff had given him a board with the alphabet written on it, and he pointed to the letters he needed to communicate with Wells. She had brought an iPod with her, and he asked for some John Prine. She played the folk singer's album "German Afternoons."

On July 27, the rest of the hikers arrived at the hospital. They were allowed into Dyer's room in small groups. Gross and Chase went in first, standing at the foot of the bed where he lay sleeping.

He was still drifting in and out of sedation. He was also having hallucinations—side effects from the medications he was taking, or from the trauma or maybe from some combination of the two. The hallucinations were nearly as terrifying as the attack itself, he said later.

As Dyer woke up, he saw his friends staring back at him. Slowly he pointed to the letters on his alphabet board and spelled out two questions: Would they all like to come to his house for a lobster bake? And would someone take him slap dancing?

It was the kind of random humor that had helped endear Dyer to the group in the first place, and seeing it now was a huge relief. Dyer really was OK. Together, they had survived the experience of a lifetime.

When the Sierra Club travellers headed into the Torngat Mountains, they hoped to see something rare—a wilderness still free from the effects of modern civilization. What they didn't know was that the modern world has already arrived in the Torngats in the form of climate change, which is altering the Arctic ecosystem right now, not in some distant scenario.

With each decade, another huge swath of the ice-covered north dissolves into the ocean, taking with it a thriving ecosystem that is uniquely adapted to survive there. As this habitat melts away, the reflective white surface of the ice gives way to dark expanses of the ocean, driving a cycle in which warming leads to melting which leads to more warming.

The evidence is in plain view in the mountains of the Torngats, newly topped with green growth, the mosquitoes buzzing nearby as the permafrost disappears and the Ski-Doos that crash through the untrustworthy ice. And during the extended warmth of summer, the evidence is in the polar bears that crowd the shore, forced to wait longer than ever to return home.

After the Torngats trip of 2013, the hikers returned to their lives—in North Carolina and New York, in New Mexico and California, in Oregon and Maine. They carried home with them not only the relentless memory of their friend's terrified screams but also the bond of having lived through an unimaginable ordeal.

Their story also points to something else, something to which they had borne witness: the advanced symptom of a much larger environmental upheaval, in which humans are forced to adjust to their own changing habitat, not just in the Arctic, but everywhere on the warming planet.

On Aug. 17, 2014, Matt Dyer emerged from the cabin of The Robert Bradford *as it headed into the mouth of Nachvak Fjord. After hours of cloudy skies and thick fog, the sky had turned a radiant blue. As the boat passed small icebergs, the sun's reflection off the ice and water caused the handful of people on deck to squint. And then they saw them: The steep peaks of the Torngat Mountains sliced down to the cold water of the fjord below.*

In his bright orange windbreaker and winter hat, Dyer walked to the side of the boat's deck, a cup of tea in his hand. He leaned against the low rail of the boat watching the scenery.

The last time he'd seen the fjord was 13 months ago, when he was loaded onto a helicopter, semiconscious and covered in blood. Now he was back, searching for closure. He wanted to experience this beautiful place on different terms, building memories of its awe-inducing splendor, rather than of the horror of a polar bear attack.

Dyer had originally planned to return to the Torngats in the summer of 2015. But when InsideClimate News and VICE Media invited him to join them in August 2014 for a week in the Torngats, he accepted the offer immediately.

This trip was much different from his first one. The group of five slept on The Robert Bradford, *not on the land. And they didn't go anywhere without Maria or Eli Merkuratsuk, the brother and sister bear guards—both armed with shotguns—who had been hired to guide and protect them.*

At a little after 4 p.m., The Robert Bradford *put down anchor in a small harbor that was separated from the site of the attack by a grassy spit of land. The clear skies and late afternoon sun gave a glow to the area. Even the weather seemed to be cooperating—the temperature was in the high 40s, far nicer than it had been around a year earlier.*

Eli Merkuratsuk found the first polar bear paw print just a minute or two after a dinghy deposited them on shore. The bear's splayed toes had pressed into the hard sand—recently.

With one bear guard up front and one at the rear, the group wove a path inland and crested a small hill. Dyer found himself looking down through binoculars at the site where he had been attacked. Within minutes, he spotted a bear. It was standing on a raised piece of land where Gross and Chase had once considered camping.

Over the next three days, the group saw eight more bears. But instead of meeting them with fear or hesitation, or backsliding into the trauma of what had happened to him, Dyer was filled with a sense of peace. Sometimes he sat quietly and stared at the landscape in an almost meditative way. Sometimes he gave in to his seemingly constant urge to make the people around him laugh—generally at his own expense.

The Merkuratsuks were a constant presence and a constant comfort, teaching as they led. On hikes they plucked edible plants for the group to taste—the Arctic tundra is covered in edible plants, yet another incongruity in this place that seems so inhospitable. They pointed to some willows and said they should be avoided, because bears tended to congregate there. Dyer said he'd used that same location as a discreet toilet a year earlier and joked that a bear could have gotten him at a particularly vulnerable moment. But it was also a sobering reminder of the many risks he and the earlier group had unknowingly taken.

Back home in Maine now, Dyer is still consumed by—maybe even obsessed with—polar bears, but not how you might expect. It's not PTSD, and he's not haunted by dreams of being mauled. He's just fascinated.

By the end of the year, he plans to get a tattoo of a polar bear on each of his forearms. If you're a tattoo guy, he explains, you don't go through an experience like that without getting a little ink. Along with the scars on his face and neck that are now covered by his newly grown ponytail and beard, and the low, husky rasp that is now his voice, the tattoos will be permanent reminders of just how close he came to death. There won't be a day in Dyer's life that he won't remember the bears.

MARTA CHASE AND RICH GROSS

Chase and Gross have been going through their own coping process in the aftermath of the attack. The Sierra Club investigated the events of the trip and found that Chase and Gross did nothing wrong. Both remain active leaders in good standing. In August, they led a trip into the Sierra Mountains that included Marilyn Frankel, Kicab Castañeda-Mendez and Matt Dyer.

Still, the long-time guides have re-examined every decision they made in those fateful days.

Both have come to the same conclusion: For the kinds of minimalist wilderness trips they lead, without armed guards, there is no certain way to stay safe in polar bear country. So neither would return to polar bear country and sleep on the land without significant changes in protective technology. It's just not worth the risk.

The Sierra Club has taken a similar stance. Tony Rango, director of the Outings Program and Program Safety for the Sierra Club, said such trips are on hold until next spring when the organization will decide whether to require armed guards on certain trips—like backpacking in polar bear country or hiking safaris in Africa where lions or other aggressive animals roam.

If they could do it over again, Chase and Gross say, they wouldn't put so much trust in that kind of electric fence.

"I would never probably go back out there again and do a backpacking trip. It's sad because it's just such an amazingly beautiful area," Chase said. "I think people need to be very much aware of the risks that they're taking, and the fact that this incident did happen."

Gross said the experience made him understand the insignificance of humans in what he calls "huge nature."

"Think of all the things that we had to do in order to survive—and just barely did," he said. "The other piece you realize is how much humans can ...destroy that [world] with global warming and carbon emissions. You realize how insignificant we are ... but also how quickly we can change that."

PARKS CANADA

Since Dyer's attack, Parks Canada has been trying to figure out what went wrong on the Sierra Club expedition and how to protect the few hundred visitors who enter the park each year from future attacks.

"We want to make sure that we're doing our due diligence, that we are providing the right messages in the right way," said Judy Rowell, superintendent of Torngat Mountains National Park. "I don't think that we've done anything wrong but there may be some things we could do better."

One step the agency expects to take next year will require any organization that leads a group into the park to apply for a business license that includes a Polar Bear Protection Plan that articulates how they will protect themselves from polar bears. The permit that Gross and Chase filed with Parks Canada included a similar question, asking what wildlife deterrents the group would take with them. But Rowell said the new requirement will go further and ask specific questions about how guides will operate in polar bear country, like where they will camp and how they will respond to a bear sighting.

At this point the agency's plans don't require visitors to travel with bear guards, Rowell said, because there are not enough guards to go around. But the possibility of requiring guards in the future "will never be off the table," she said.

Rowell said a number of missteps added up to one catastrophe.

One of the group's biggest mistakes was setting up camp near Nachvak Fjord. The area is known "a polar bear highway," she said, and Parks Canada and Base Camp staff see bears there regularly.

"You wouldn't have seen any bear guard agreeing to camp there," she said. "Our folks here would go a minimum of 10 kilometers inland. And even then you need to be careful, very careful. When we say stay away from the coast we mean 10 kilometers inland."

Gross and Chase say that in all of their conversations and emails with Parks Canada, including the vetting of their hiking route, no one ever mentioned that the area was a "polar bear highway" or that they should hike 10 kilometers inland before setting up camp.

That point—about moving 10 kilometers inland—isn't articulated in the polar bear safety brochure that Parks Canada sent to the group. Nor is it mentioned in the polar bear safety video Parks Canada produced.

Rowell also said that after the encounter with the bear that was watching them, they should have set up a 24-hour watch and then relocated. "If I was in an area where I saw a bear hanging around I wouldn't be very relaxed bedding down at night out without someone keeping an eye on where it was going or what it was doing," she said.

One obvious response could be to make bear guards mandatory—to require that all groups enlist the services of a trained, armed Inuit bear guard. But Rowell said that while they continue to talk about it, it's not on the table for now. They can't require a bear guard without having a supply of guards available for anyone interested in going into the park, she said, and right now, they don't have enough people who want to do the job. But "it will never be off the table," she said.

The summer of 2014 offered more evidence that the bears are coming on land and interacting more with humans.

In late June, before Base Camp's formidable electric fence was assembled, a polar bear wandered into camp. Base Camp manager

Wayne Broomfield told the CBC that he had gone outside to try to scare the bear away when it tried to attack him. It chased him back into one of the Base Camp buildings, biting and scratching at the door when he slammed it shut. It was a big bear, but also skinny, Broomfield told the CBC. It was also more aggressive and determined than most bears. Eventually, they chased it away with a helicopter.

THE BEARS

It's impossible to get inside the head of a polar bear. There's no way to tell exactly why the bear attacked, or why it chose Dyer over some other member of the group, although some suspect that it was drawn by Dyer's impressive snoring.

But some things are knowable, and in the aftermath of the attack Parks Canada conducted an investigation to determine what they could about what happened and why.

By looking at the paw prints and behavior of the bear after the fact, Parks Canada determined that the bear that attacked Dyer came from just west of where they were camping, an area between the campsite and the ledge where the bear had been watching them the day before the attack. At the edge of the beach, which separated the campsite from the hill and ledge where the bear had watched them, there are thick shrubs and willows. Perfect for a bear to hide in.

"At some point the bear made a direct beeline towards the camp to the tent where the victim was camping," said Peter Deering, Parks Canada's resource conservation manager for western Newfoundland and Labrador. "It's not like he came and spent some time poking around the periphery.

"It appears that it was a very focused attack."

Since the bear wasn't killed—and all parties involved with the incident are glad that it wasn't—there's also no way to determine what might have led it to attack a human. It could have had a bad tooth or an injury that had prevented it from eating, leaving it hungry. It could have been a curious, young bear that wanted to taste whatever it was that was sleeping in its territory. Or it could have been left hungry by a shortened season on the ice, seeking something to fill its stomach until it could get back onto the ice and hunt seals.

Given the circumstances of the group's trip—where they were camping and when, and the fact that there was a bear that was clearly determined to attack—Deering said there's a certain inevitability to what happened.

"Without a guard posted, without an alarm going off—I'm not sure anything could have stopped this."

ABOUT THE AUTHOR

Sabrina Shankman is a producer and reporter for InsideClimate News. She joined InsideClimate in the fall of 2013, after helping produce documentaries and interactives for the PBS show "Frontline" since 2010 with 2over10 Media. In 2012, she produced the online interactive *A Perfect Terrorist: David Coleman Headley's Web of Betrayal,* which won an Overseas Press Club of America award.

Prior to working with Frontline, Shankman reported for ProPublica, the Wall Street Journal and the Associated Press. Her work has been honored by the Society of Professional Journalists and the Society of Environmental Journalists, and she was named a finalist for the Livingston Awards for Young Journalists in 2010. Shankman has a Masters in Journalism from UC Berkeley's Graduate School of Journalism.

InsideClimate News is a Pulitzer prize-winning, non-profit, non-partisan news organization that covers clean energy, carbon energy, nuclear energy and environmental science—plus the territory in between where law, policy and public opinion are shaped.

Made in the USA
San Bernardino, CA
15 July 2015